NATIONAL GEOGRAPHIC

School Publishing

Animals *of* Denali

PIONEER EDITION

By Susan E. Goodman

CONTENTS

Denali National Park in Alaska is a harsh land. It's too cold for most people. But for many plants and animals, Denali is a paradise.

Animal
De

A brown bear and her cub

*f*nali

By Susan E. Goodman

Red fox

Caribou

Great
horned owl

A grizzly bear walks out of her den. She is slow and clumsy. She has been sleeping all winter long.

The bear is not alone. Her cubs follow her. She gave birth to them in her sleep.

The bears are lucky. They slept through Alaska's long, harsh winter. Other animals struggled in the cold.

Harsh Winter

Denali National Park has tall mountains and thick forests. In some places, sheets of ice cover the land. The park was created in 1917. It was made to protect the land and wildlife of Alaska.

Much of the park's land is **tundra**. It is open land with grasses and shrubs. The tundra has no trees. Winter is very cold. There are only four or five hours of sunlight each day. Snow covers the land.

Alaska's animals need lots of land. For example, a single pack of wolves uses 389 square kilometers (150 square miles) of land for hunting.

Staying Behind

Only a few kinds of animals spend winter in Denali. Moose and wolves do. But staying alive is not easy. Food is hard to find in the deep snow.

Some animals, such as bears, **hibernate**. They go into a deep winter sleep. They do not eat. They stay in their dens, or underground homes. They sleep until spring.

Pack Animal. Gray wolves often travel in packs. Parents and pups usually form a pack. BELOW: Caribou migrate across the tundra. BELOW RIGHT: Owls hunt at night.

On the Move

Many animals leave for the winter. They **migrate**, or travel, south. The weather is warmer there. Food is much easier to find.

Caribou, or reindeer, migrate each winter. They travel many miles across the tundra. Their flat hooves help them walk across icy or wet ground.

In spring, warmer temperatures return to Denali. So do many plants and animals.

Spring Blossoms

After many months, Denali's winter changes into spring. Warm weather returns. The tundra comes alive. Flowers cover the ground. Animals shed their winter coats. Some even change color.

The snowshoe hare had a white coat in winter. White fur was hard to see in the snow. In the spring, the hare grows brown fur. It blends in with the ground. The hare's fur helps it hide from **predators**, or animals that eat other animals.

Summer Visitors

In summer, the tundra fills with animals. Birds fly in from all over the world. Herds of caribou come back from their winter homes.

Food is easier to find. Moose and caribou eat plants. Grizzly bears snack on roots and blueberries. Owls swoop down from the sky to hunt. They eat animals, such as mice and hares.

Summer also brings many insects. In fact, more mosquitoes live here than anywhere else. A caribou can lose a lot of blood each week from bug bites!

The Big Chill

At the end of summer, days begin to get cooler. Animals get ready for winter. Dall sheep grow thicker coats. Grizzlies start eating— a lot. They gain many pounds each week.

The tundra's summer visitors start to leave. They head to their winter homes. Caribou go back to the forests.

The arctic tern has much farther to go. This bird migrates to Antarctica. That is more than 32,187 kilometers (20,000 miles) away.

Deep Freeze

By November, the weather is cold again. Snow falls. Food is much harder to find.

Grizzly bears stop eating. Soon they will crawl inside their dens. They will sleep through the winter.

Snow covers the ground. The tundra seems still and lifeless. A hard winter lies ahead, but one day spring will return. The cycle of life on the tundra will start again.

High Point. Wind blows snow off Mount McKinley in Denali National Park. ABOVE RIGHT: Dall sheep live on mountains in Denali.

Wordwise

hibernate: to sleep deeply during winter

migrate: to move

predator: animal that eats other animals

tundra: cold, treeless plain

The Making of a Park

Alone in the Snow. Denali
National Park is a perfect place
to enjoy nature. Yet it took many
people to protect this land.

Alaska is not the easiest place to live. During winter, it is freezing cold. Food is scarce and hard to find. Yet many animals thrive in Alaska's wilderness. They are free to roam from place to place. Denali National Park was made to protect this wild land.

Landing on Water. Parts of Alaska have few roads. So people often use seaplanes to travel through the state.

Dreaming of a Park

Denali National Park is one of the largest parks in the United States. It covers more than 2.4 million hectares (6 million acres). Denali was the first national park in Alaska. It started as one man's dream.

In the early 1900s, a man named Charles Sheldon dreamed of a park. Animals could live wild and free. People could visit. They could see wildlife in a wild land.

At the time, there were only a few national parks. Sheldon hoped to have another one in Alaska. Yet making the park was not easy.

Struggle Over Land

Not everyone liked Sheldon's idea. Many people needed the land.

Miners dug for gold, silver, copper, and lead. Other people hunted animals for a living. These people worried that the park might cost them their jobs. They might lose their way of life.

However, many other people liked Sheldon's idea. They wanted a park to protect the animals in Alaska. People also wanted to save Alaska's land.

Hiding Out. This snowshoe hare lives in Denali all year.

Hiking Haven. Denali has many trails that let hikers enjoy nature.

Wild Land. Denali protects a large area of land for wildlife. Caribou and other animals can roam for miles in the park.

People for the Park

In 1917, NATIONAL GEOGRAPHIC magazine printed an article about Denali. It said that many animals would be killed off without a park.

People across the country took action. They wrote to Congress. Congress listened. Soon the country had a new national park.

Bigger and Better

Over time, Denali National Park has grown. In 1980, millions of acres were added. Now the park is four times its original size!

Charles Sheldon would be happy to see Denali today. His dream came true. Denali is a place for animals— and for people too.

Life in Denali

It's time to discover how much you've learned about Denali.

1 What is winter like in Denali?

2 Why do some animals migrate?

3 What changes take place in Denali as spring arrives?

4 How did Charles Sheldon help create Denali National Park?

5 Why did people disagree about making Denali National Park?